GW00726938

MARS EXPLORER ID

YOUR
PHOTO
HERE

NAME:

FREE TEACHING RESOURCES

Download comprehensive teaching notes, curriculum guides and multimedia for each activity in this book at **MARSDIARY.ORG**.

The Mission Mars Diary is a creative STEM resource for UK primary schools. It is designed to engage students in science by learning about real-life STEM projects that are exploring ways to get humans to Mars, in particular the UK's human and robotics exploration programmes.

Curved House Kids and Lucy Hawking are extremely grateful for the support of the UK Space Agency, without whom this book and the accompanying teaching resources would not be possible. We would also like to thank Pamela Burnard, Professor of Arts, Creativities and Educations at the Univeristy of Cambridge and Peter McOwan, Professor of Computer Science and Vice Principal (Public Engagement and Student Enterprise) at Queen Mary University of London, for academic input and for reviewing the materials.

First published 2018
by Curved House Kids Ltd
60 Farringdon Road
London EC1R 3GA
www.curvedhousekids.com
info@curvedhousekids.com

Published by Kristen Harrison
Written by Lucy Hawking
  with Kristen Harrison and Lucie Stevens
Illustrated by Ben Hawkes
Designed by Alice Connew
Project coordination by Lucie Stevens
Digital management by Alice Connew

The right of Curved House Kids Ltd and Lucy Hawking to be identified as authors of this work has been asserted in accordance with Section 77 of the Copyright, Designs and Patents Act 1988.

A CIP record for this book is available from the British Library.
ISBN 978-0-9927302-8-4

Printed in the United Kingdom, on sustainable paper, by Newnorth Print.

Acknowledgements
The Mission Mars Diary team would like to thank everyone who has helped with the development of this programme, especially Susan Buckle, Sue Horne, Libby Jackson and Rachel Luke at the UK Space Agency; ESA astronaut Tim Peake and the European Space Agency; our amazing illustrator Ben Hawkes; Professor Peter McOwan and Professor Pamela Burnard for academic input; Tom Lyons and colleagues at the National STEM Learning Centre; Hannah Coulson for the JWST illustration; Cindy Forde and all of the amazing experts who have volunteered their time to advise us on everything from Martian volcanoes to engineering a Mars rover: Dr Maggie Aderin-Pocock, Dr Abbie Hutty, Professor Stephen Lewis, Vinita Marwaha Madill and Professor Tamsin Mather. Thanks also to Jan Boström at JGB Service for the Maze Generator.

Finally, we are indebted to our incredible team of teachers who have tirelessly reviewed and updated the teaching materials and given us invaluable feedback: Paul Cameron, Hannah Chivers, Laura Cowan, Ceri DeRoy-Jones, Claire Loizos, Deirdre Mulrooney and Nicola Sivier.

Your Mission MARS DIARY

Written by Lucy Hawking,
the Space Crew and YOU!

Illustrated by Ben Hawkes

THE CURVED HOUSE *kids*

WELCOME, MARS EXPLORER!

Wow, you are brave! You're going to Mars to build a new habitat for humankind. You will need to design a rocket, choose your crew, explore the surface of Mars, do experiments in space, communicate with Earth and so much more. This is a big task but we know you can do it. Thanks, by the way, for agreeing to venture further into space than any human has gone before. Everyone on Earth is cheering you on!

Before you blast off on your adventure, you've got some planning to do. What tasks will you do, and which ones will you give to your robotic friends? How will you get safely to the Red Planet and what will you eat on the way? What will you build on Mars? What kind of society would you like to create and who will live there? Big questions, our intrepid friends, so go forth and find the answers...

Good luck – and don't forget us!

Lucy Hawking
and the Space Crew

THIS WAY TO MARS!

SPACE WORDS

Keep your eyes peeled for new words as you work through your Mars Diary. Find out what they mean and add them to the **Space Glossary** at the back of the book. Here are some words to get you started, can you find out what they mean?

Alien A life form that exists outside Earth. Scientists believe that, in the past, the environment on Mars could have supported the evolution of life but this would have been no more than tiny microbes!

ExoMars

Methane

Orbiter

Robot

CHAPTER ONE:
LIFE ON MARS

Why do we send humans and robots into space?
What will Mars be like, compared to Earth? Does
anything live on Mars already – and if so, what does
it look like? Complete the challenges in this chapter
to have your alien life detectors at the ready for your
adventure on the Red Planet...

SIGNS OF LIFE

Zap to discover more about Earth and Mars

EARTH

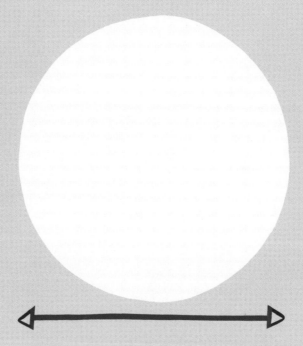

Diameter: 12,742 kilometres

If an alien looked at Earth from far away, how would it know there is life here? Draw and label Earth and Mars and compare the differences. What are the signs of life on each of these planets?

MARS

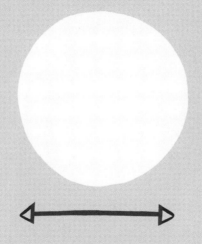

Diameter: _____ (approximate)

Hi explorers,
Mars is *approximately** half the size of Earth. Can you work out the diameter of Mars?

*Add this word to your Space Glossary!

MAKING HISTORY

Zap to travel through space history

Before you set off on your epic journey to Mars, find out about some of the amazing human and robotic missions from the past, present and future. Can you find out who went WHERE, WHEN and WHY?

Neil Armstrong and Buzz Aldrin

Yuri Gagarin
The Russian Vostok 1 mission in 1961 made Yuri the first human in space!

Tim Peake and the ISS

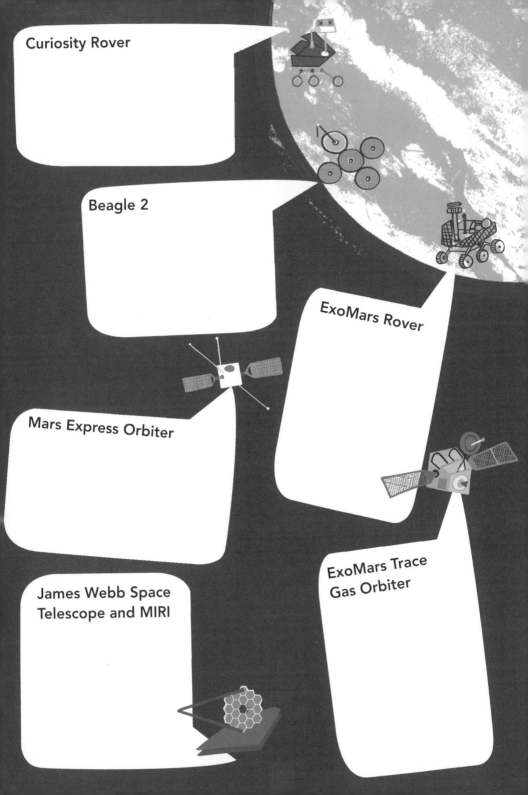

Curiosity Rover

Beagle 2

ExoMars Rover

Mars Express Orbiter

James Webb Space Telescope and MIRI

ExoMars Trace Gas Orbiter

MARS QUIZ

Zap for the answers!

Hi explorers,

I'm Sue Horne, Head of Space Exploration at the UK Space Agency. Let's test your knowledge of the Red Planet before we send you on this important mission.

For number 10, create your own question and test your friends!

TRUE OR FALSE?
YOU TELL US...

	True	False
1. The planet Mars was named after the Mars Bar.	☐	☐
2. Lower gravity on Mars means you would bounce twice as high as on Earth.	☐	☐
3. Mars is the eighth planet from the sun.	☐	☐
4. Because Mars has an atmosphere, it also has weather!	☐	☐
5. A year on Mars is 1000 Earth days long.	☐	☐
6. Mars has the highest mountain in the solar system, Olympus Mons.	☐	☐
7. Mars has two moons.	☐	☐
8. Humans have successfully flown to Mars and founded a Martian city.	☐	☐
9. Mars is visible from Earth with the naked eye!	☐	☐
10.	☐	☐

LETTER TO TIM

Zap to watch
Tim's message!

Whoa, you've received a video message from ESA
astronaut Tim Peake! He wants to know why you're
so keen to get to Mars. Zap to watch, then write a
reply explaining the reasons for your mission.

WORD SEARCH 1

Zap for the answers!

Find the words you've learnt in this chapter and add them to your Space Glossary at the back of the book. Words can go in any direction.

X	R	I	I	E	O	V	X	V	G
U	E	Z	Q	R	C	E	C	R	R
W	I	F	K	E	A	Y	A	E	A
K	U	D	W	H	A	B	S	V	V
H	G	A	Z	P	O	B	T	O	I
T	B	Q	N	S	R	N	R	R	T
R	P	K	S	O	B	I	O	Z	Y
A	Q	R	D	M	I	U	N	Z	Y
E	A	P	K	T	T	L	A	Z	O
M	Q	M	R	A	E	E	U	Z	I
N	E	I	L	A	R	T	T	L	C

Target = 8 words beginning with:

A E O A G R A M

CHAPTER TWO:
PLAN YOUR MISSION

Find out what you need to know about your destination and decide who's coming with you from planet Earth! Pack a space suitcase and jump on board your spacecraft when you have made sure your mission planning is the best in the solar system.

GOING THE DISTANCE

You've got a long journey ahead. Let's work out how far it is from Earth to Mars. But wait, the distance keeps on changing! Can you find out why?

Clue: Earth = 1.0 AU or 150 million kilometres from the sun

Neptune

Saturn

Uranus

Jupiter

Venus

Mars

Mercury

Earth

Sun

It takes between 150 and 300 days to get to Mars from Earth! Why does the journey time change like this? And when is the best time to launch your rocket to Mars?

Zap to learn about orbits

Scientists measure the distance from the sun in Astronomical Units (AU). The clue on the left gives you the distance when Earth and Mars are closest together. If Earth is 1.0 AU from the sun and Mars is approximately 1.5 AU from the sun, what is the distance between Earth and Mars?

.... AU   or .... million kilometres

ASTRONAUTS WANTED!

Time to decide who will be part of your Mission Mars crew. You can take two people – choose wisely! Let's start the recruitment process...

STEP 1:
What skills and qualities are you looking for in your crewmates? Write your ideas here.

Zap to meet real space experts

STEP 2: Plan the interviews. What questions will you ask candidates to find out if they have what it takes to join your team?

Question 1:

Question 2:

Question 3:

STEP 3: Interview your friends and pick two crewmates. Draw and name them here!

EXCESS BAGGAGE

You have your crew, now it's time to pack. Everything you need to live and work on the Red Planet needs to squeeze into your space suitcase. What will you take?

CHECKLIST:

- [ ] Clothing & toiletries
- [ ] Toolkit
- [ ] Emergency gear
- [ ] Food & drink
- [ ] Entertainment
- [ ] Communications
- [ ] A keepsake

Zap to find out what keepsake Tim Peake took to space!

Draw your items here

DESIGN YOUR OWN ROCKET

Hi space explorers!
I'm Vinita Marwaha Madill, a space engineer. Draw and label your Mars rocket here and use the checklist so you don't forget the essentials!

CHECKLIST

☐ communications

☐ safety

☐ propulsion

☐ fuel

☐ space for the crew!

Zap for
inspiration!

WORD SEARCH 2

Zap for the answers!

Find the words you've learnt in this chapter and add them to your Space Glossary at the back of the book. Words can go in any direction.

P	Y	Y	K	L	P	T	J	Y	T
V	R	C	O	M	V	H	G	R	I
O	E	O	N	C	R	E	W	U	U
Z	D	H	P	E	O	C	Y	C	R
S	V	W	I	U	G	C	G	R	C
E	A	E	J	C	L	R	C	E	E
Z	D	F	T	M	L	S	E	M	R
L	E	Q	E	U	T	E	I	M	S
F	U	E	L	T	T	K	C	O	E
S	U	N	E	V	Y	I	F	R	N
T	I	B	E	I	U	V	A	Q	L

Target = 9 words beginning with:

C E F M P R S V V

CHAPTER THREE: YOUR NEW HOME

Time to get out of your spacecraft and stretch your legs after the long journey! Take a look around you and record your findings in your Martian log book. But watch out! There are dust devils around...

WEATHER ON MARS

The weather on Mars is very different to Earth, so you'd better be prepared! Create a weather report showing a day on Mars and make sure it's colourful and visual to help your crew understand the conditions they'll be facing.

Hi explorers,

I'm Stephen Lewis and I study the weather on Mars. I've collected 24 hours of data from the Curiosity rover on Mars. Use it to create an infographic.
Can you include some comparisons to the weather on Earth?

Zap for
Mars weather
data

BREAKING NEWS

Scientists on Earth have been observing your new home and have made a Mars-shattering discovery! Satellite images appear to show black rivers on Mars. Can you investigate and report back with a news story?

Zap to view images from the Mars Express orbiter.

Hi Mars explorers!
I'm Cindy Forde, a science communicator. You'll need a catchy headline and a great image to get our attention. Make sure your story is based on the most up-to-date science and facts too.

MARS DAILY

DATE:

MIGHTY MONS

The surface of Mars is peppered with mountains, some of which appear to be volcanoes. Mars orbiters have captured images of a huge volcano called Olympus Mons. Now that you're on the Martian surface we need you to investigate.

Hi Martians,
I'm Tamsin Mather, a volcanologist on Earth. Can you help us picture what this giant volcano might look like from the surface of Mars? How tall is it and how does it compare to volcanoes on Earth? Draw and label a diagram to show us what you can see from the ground!

Zap to activate your Mars cameras